SMOKE, ASH AND BURNING EMBERS

A Handbook for Cigar Play by

John D. Weal

Published by The Nazca Plains Corporation
Las Vegas, Nevada
2010

ISBN: 978-1-935509-72-1

Published by

The Nazca Plains Corporation ®
4640 Paradise Rd, Suite 141
Las Vegas NV 89109-8000

Photographer, Chris Landers
Art Director, Blake Stephens

DEDICATION

This book is dedicated to boy keith and boy andrew, the two boys who modeled in this book.

First boy keith, as he has come into my life and through his daily service and dedication has changed my life. He also submitted to me without any reservations on many things he didn't know anything about, when he first met me, and blindly trusted me when he had reached a low point in his life. Through that trust, I hope to show him that because of that trust, he will truly be rewarded in life. He has earned a special place in my life and hopefully he will understand that through his life in service, that after his permanent collar, which will be soon, he will accept his new life through his marriage to me, as I have proposed to him. He tearfully accepted.

boy andrew who opened up his life in service to me on his visit for the photo shoot and became special to me through his background and family ties to the Holocaust.

SPECIAL THANKS

A special thank you is given to the following for allowing and being part of this journey with me.

Chris Landers - a very dear friend, photographer for this book and

In service to Miss Rae, Ft Lauderdale, Fl.

My friend Pup (Lorenzo), for helping with the dual gas masks scene.

Master Oakman and slave diamond, for the use of their outdoor dungeon space, Ft Lauderdale, Fl.

Slave George (Jorge Pena), for his service, friendship and for always being there whenever I have needed something in my life.

Zak and Steve, for the outside use of the patio at the Ramrod Bar, Ft Lauderdale, Fl.

Jeffrey Payne, IML 2009, David Roy, his partner, and boy adam for being true friends.

Allen Roe, for being a lifelong friend and being there when I needed him through out my life.

LeatherWerks, Ft Lauderdale, Fl., for being involved and my friends in my leather journey.

A SPECIAL REMEMBERANCE:

TO MASTER RICHARD

1941-1979

In loving memory

Sir, I will always be in service to you. You walk with me daily through this journey of life. I will always remember our days together in the Castro and because of you, and what we experienced, I live to make you proud!

This book celebrates our lives!

I hope this book will save someone from being hurt, which was our dream; to educate!

TABLE OF CONTENTS

PURPOSE OF THE BOOK

There are many thoughts that come to one's mind when the idea of cigar play in leather is mentioned. In these pages I will try to give you a good overall view of how to play and what can be done to heighten sexual excitement, while still being careful and safe in play between two consenting adults. This book is aimed at male to male play, however; all other gender types will be able to benefit from this book, as the play is the same, just maybe the body part may be different, i.e.: instead of a cock you could use the vaginal area, but an ass is an ass, no matter what the gender.

This book is not intended to make you a cigar aficionado as there are books, magazines and web sites out on that subject. They update constantly with views of the market place and specials to buy cigar products etc. I will, however, talk about some basic cigar history, overall use, storage and lighting of a cigar prior to talking about playing with a bottom in many types of cigar play. This book hopefully will open your mind to a subject matter of SM sex that will stimulate you to become more creative with a cigar than just to sit back, relax and enjoy it.

Hopefully, you will learn the process and techniques so you can expand every subject matter and heighten both you and your bottoms ability to enjoy cigar smoke, ash, and eventually a small burn. Cigars are making a huge comeback and have increased in sales by 28% in the early 2000 years. There are many websites which will educate you on buying cigars and there is even an encyclopedia on cigars but there is nothing that I could find where it teaches you how to play with cigars. Thus the purpose of my book and photos to help you enjoy and see what I am trying to convey to you. I hope you will enjoy these pages and pictures. I hope you will enjoy my style of writing. This is the first in a series of approximately ten books I will be writing and documenting my play and life. Upon finishing this book, I hope you will find a way to incorporate a cigar into your SM scene.

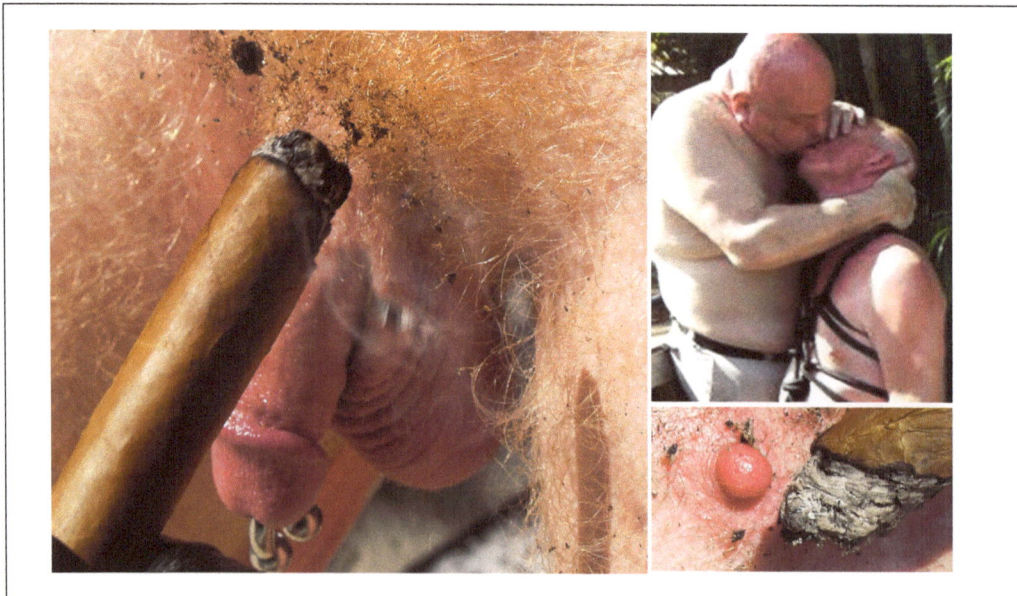

HISTORY OF CIGARS

The first record of anyone using tobacco is thought to be that of the Mayas of Central America, which historians believe were smoking as early as 1000 B.C. The first known discovery of tobacco, by Europeans, was during Christopher Columbus' discovery of North America in 1492 and the meetings with the Indians. A discovery that changed the world for everyone and thus the history of the cigar started. Unlike cigars of today, they were smoking a plant which became known as tobacco, twisted into plantain leaves, palm leaves or corn husks when smoked producing an unusual fragrance.

The first to use tobacco from Columbus's expeditions are said to be two of Columbus's crewmen known as Rodrigo de Jerez and Luis Torres. Upon returning from the expedition, they took some tobacco back to Spain where it is recorded that Jerez was seen smoking tobacco. Locals thought he had been possessed during his voyage with Columbus when they saw smoke coming from his mouth. The locals immediately turned him into the church authorities where he was tried, convicted and was put in prison. Tobacco usage did not stop there. It became very common for many of the men on the ship and it soon became a popular pastime. By the time Jerez was released, tobacco was commonly used throughout Spain. It eventually spread to Portugal, France, Italy, Great Britain and eventually made its way back to North American colonies. It took nearly 200 years to spread throughout all of Europe.

In the early 1600's, tobacco started to be grown in abundance and in turn became a commercial product for farmers. During these early years, tobacco was thought to have medicinal qualities. Phillip II of Spain and James I of England were two of the most notorious detractors of tobacco during this period. However, despite the objections, it gained favor with the locals of Great Britain and was smoked predominantly in pipes.

Spain during this period became more proficient in the growing and the drying of tobacco. They also found more user friendly ways of smoking the tobacco, such as rolling the leaves in tight rolls and thus giving birth to the cigar as we know it today. The rolling of the tobacco made it more manageable and smaller, with no need for a pipe. It is because of this that Spain is attributed with naming the cigar. The Mayan words for cigar were "Siyas", "Sikar", "Segars", and the Spanish word is "Ciggaro". Thus we have the modern word cigar coming from the Spanish word.

Today it takes between two to three years from the tobacco seed to the cigar, finished and ready for smoking. It grows for 18 weeks and then is fermented for six weeks and aged for approximately 18 to 24 months.

INTRODUCTION TO CIGARS

Cigar smoking means different things to different people. There are those who consider cigar smoking a private time to enjoy the taste and the aroma. Other options for smoking a cigar are socializing with friends or sharing some coffee or liquors over a cigar with your loved ones. Still others smoke cigars in celebrations like the birth of a new baby, a wedding, a holy union or an anniversary, a milestone in one's life. But today cigar smoking is what you want it to be.

In olden days, it was thought that only higher class people were allowed to purchase and smoke cigars and that only the rich could afford them. Today, everyone can afford them. Priced from a $50 brand to some off brand for a few dollars, anyone can enjoy a cigar. Different people have their own take on a cigar smoking time. What it means to you as well as how much a good cigar should cost is totally up to the individual. Cigar smoking can be calming and therapeutic and can actually cause your blood pressure to drop because you are relaxing when you smoke a cigar. It is said by most cigar smokers, that a cigar smoking time is a feeling, a joy like no other, and each person has their own take as to what it means to them. With the addition of a good coffee or tea, a scotch whiskey, good bourbon, cognac or your drink of choice, it adds up to a very relaxing time where you can savor the flavor of the cigar and your choice of alcohol. Again, depending on the person smoking, this can come at any time, but is mostly done following a good meal or dinner, when your belly is full and it is time to have a little adult enjoyment and relaxation.

Remember, when choosing a cigar, the price tag doesn't mean it is any better tasting or better wrapped. It is all about the marketing and flavor additives. If you go to any good cigar store and want to start smoking a cigar, talk to the tobacconist behind the counter, tell him or her the flavors you enjoy and the size you think you want and they should be able to recommend something until you find a favorite. Try new cigars and keep trying cigars as they are constantly coming out with new flavors.

We need to discuss some other terms which make up a cigar, such as some types of leaves which you will certainly come in contact with. Two of the more popular leaves are Maduro and Connecticut leaves. These are used to wrap the cigar and are the outside leaf of the cigar or the wrapper. A cap is a smaller leaf, which is cut and put on the end of a cigar to hold the wrap together. This is also known as the head of the cigar or the end you will smoke. You should always break the head when smoking the cigar to allow the air to flow through the cigar, causing a good even burn. This is usually done with a cigar cutter or punch but some men will put a small bite into the end of the cigar and spit it out thinking its more butch. This is not recommended, however, as it will cause the cigar to unwrap during smoking and you will get bits and pieces of the leaves in your mouth.

The end that is lit is known as the foot of a cigar. An easy way to remember this, the head of a cigar is like the head of a penis, it goes into your mouth when giving oral sex. Many men do not like to suck on the foot therefore, the foot is the part that is lit or not in your mouth.

The length of the cigar is known as the barrel of the cigar. The flavorings and types of tobacco which make up the cigar are known as the blend. Cigars come in various lengths with the smallest being a Petit, Cigarillo, Corona, Lonsdale, Robusto, Churchill and the largest being a Torpedo. Remember, this is the length of the cigar and the length of a cigar is measured in inches.

The width of a cigar is measured according to its ring size or ring gauge. Ring gauge simply measures a cigars thickness, the larger or thicker the ring gauge, the fuller the flavor. In technical terms, it's a measure of a cigar's diameter where one "ring" equals 1/64th of an inch. So, a 48 ring gauge is 48/64, or 3/4, inch in diameter. I personally like big thick cigars, usually with a ring gauge of 54 or larger torpedo length, for smoking pleasure and for a Master type of look in leather gear. I generally find anything between 40-54 gauge ring is a good cigar size roundness for play and allows a nice thick ash.

The chart below will help you visualize the most common cigar rings gauges used in manufacturing.

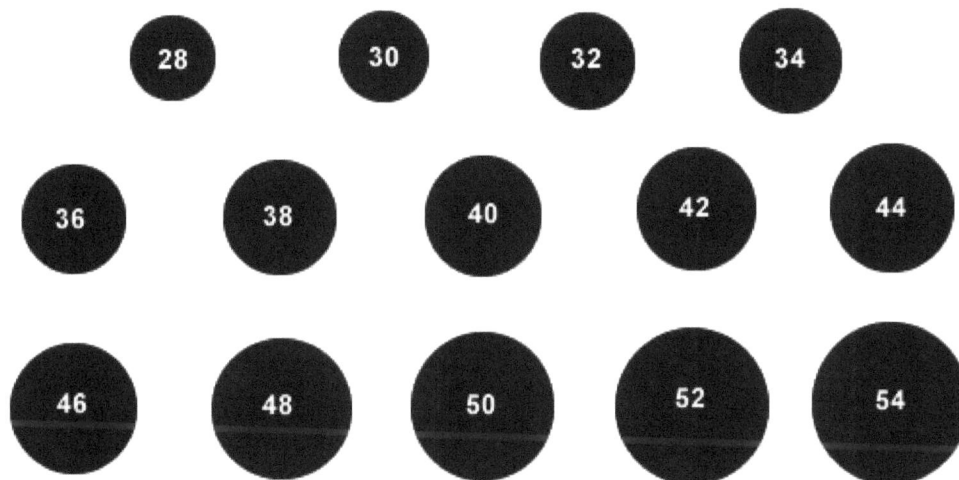

28 30 32 34

36 38 40 42 44

46 48 50 52 54

Today many countries produce tobacco for cigars. Some of the more famous growers are Brazil, Cuba, Honduras, Dominican Republic, Ecuador, Nicaragua, Cameroon (a central African Republic), Mexico. Philippines, Indonesia, Spain and, of course, the United States. Delaware, Connecticut, Pennsylvania, Maryland, Kentucky, North Carolina and South Carolina are the primary commercial growers today, but other states do grow tobacco for local use and produce some great cigars. You will be amazed at the information available on this subject if you ever do any research on tobacco growing in the U.S.

I must state that cigar smoking can be hazardous to your health as a legal disclaimer.

WHY SMOKE A CIGAR

First, let's talk about cigars overall. Many people associate cigars with an authoritative figure, as many of our fathers, grandfathers, or other important people in our lives smoked them as we grew up. Some people are mentally and sexually turned on by the smell of a good cigar as it represents these types of people or authority and achievement. Many Masters, Sirs, Daddy's and, yes, even boys/slaves/submissives love to smell a cigar, smoke one, or be the recipient of the hot sexual play. There is something about the smoke running over your tongue or your body that can be very arousing. To know the power it represents and to know what can be the result of the play would or could send many people into an erotic endorphin arousal. Also, knowing that the top can hurt you with the cigar is a sense of submission that is acknowledge by the bottom in cigar play. This is where the power side of the play comes into place. Whether you are tied up in bondage when the play is being done or just kneeling naked, servicing the top knowing he can drop ashes into your mouth or down your back as you please him can be that of utter excitement or fear. We will get into these and many other types of play in the pages to follow, but rest assured, I enjoy playing with cigars, smoking them, and watching what happens to people when I smoke them. I hope you will enjoy this book and take from it the years of experience and enjoyment I have in smoking a cigar as well as playing with it. I also hope it will help you to become a good cigar smoking top or a bottom who now understands why we play with cigars and their smoke.

HUMIDORS

Storage of cigars is something you have to understand. Cigars should be well conditioned before smoking. Cigars should be aged for several months to several years at the proper humidity (70% Relative Humidity) and temperature (70° F). A dry cigar will burn hot, fast and can taste harsh. A damp cigar will be hard to light and can be hard to draw. A dried out cigar is not a good burning cigar but a hazard in this play. I keep my cigars well moistened in a humidor.

There are many humidors that can be purchased, from a simple small desk or counter top model to an elaborate one that will take up a room in your house. Mine is a simple desktop model with separators and a hygrometer measuring the moisture in the humidor. Any humidor I would recommend you buy has not only a water reservoir but also a hygrometer that will measure the humidity in the humidor. These are very important factors in maintaining a good cigar that will burn well and not flake its leaves during play, causing a hazard or possible third degree burns. On the other hand, some people do enjoy burning and we will discuss this later in the book.

The most crucial characteristic of a fine humidor, however, is that it can provide a consistently tropical environment. A constant humidity level will also increase your smoking pleasure, beyond providing proper burning speed, with less heat and therefore more flavor. A humidor should always be measured against its ability to provide constant humidity to its contents over a long period of time. Remember, this does not only mean how often you add water to the humidification system; it also means that 20 years from now the box lid hasn't warped and the hinges still open easily and quietly. A humidor lid should not be airtight, to allow the necessary circulation of air. Musty smells destroy cigars. Don't forget about humidity: most humidifiers rely on some variety of sponge or chemical compound. Whatever the medium, remember that prime cigar aging demands constant humidity levels. Always place dry cigars on the bottom of the box first, where they will regain humidity slowly, then move them to the top shelf. Analog hygrometers, while fancy looking, are seldom accurate even in the most expensive desktop models.

Cigars are a natural agricultural product, and the quality of the smoke depends largely upon their care by the user. The best made cigars in the world - constructed of the finest leaves by the most experienced rollers - will mean nothing if they are not maintained properly.

CUTTERS, LIGHTERS, AND ASH TRAYS

When smoking cigars you should have a cigar cutter and a cigar lighter. Again, cigar cutters come in a variety of styles and a variety of prices. There is a simple blade guillotine, a traditional cut that lowers down over the end of the cigar. Then you have a very elaborate scissor type (V cut). This V cut is usually made of silver or gold and will cut the end off with a v shape. Thirdly, a cigar punch. All cuts should be made quickly and decisively. The cut should be made just above the cigars cap line (the curved area that covers the head of the cigar). The cut allows the cigar to be readily lit.

The traditional cut takes a straight slice across the cigars head/cap line. This is the best cut to create an easy well circulated draw. However, this cut does allow for residue and tar from the burning tobacco to come in direct contact with the smokers mouth.

The V cut creates a wedge shape notch in the cigars cap. This cut allows proper air circulation to occur. The smokes tar and residue accumulate on the sides of the wedge keeping the bitter taste away from the smoker's mouth. It is difficult however, to keep a V cutter sharp because of its unique shape and cutting process. It is also hard to maintain and keep the v shape on the cigar during smoking for optimum smoking.

Then we have the punch style, which is like a construction punch, you literally punch a hole in the head/cap of the cigar. Removing the punch and the tobacco creates a passageway for the smoke and air to flow. Depending on the diameter of the punch, air circulation may be restricted and the smokes tar and residue can accumulate around the opening.

What a good cutter does for the cigar is basically give you a good clean end to inhale the smoke through the cigar, getting all the aromas of the cigar and allowing you to enjoy the taste, smell and feelings a cigar can give.

A cigar lighter is different from a cigarette lighter as it burns more intensely, which makes for a more even lighting of the cigar, making it much easier to light. I recommend you buy a four burner butane lighter, if you can find it, but a three burner works just as well. It will light the cigar quickly and evenly but be careful of the intense flame. Unlike cigarettes, where the leaves are chopped very small and wrapped in paper, a cigar is wrapped in cigar leaves. These leaves, depending on how tightly they are wrapped and packed, make it harder to light. Therefore, the intensity of the lighter makes for a good burn and a good even light.

Ashtrays for cigars are made to keep them from rolling and keep the glowing embers in the center of the ashtray. If you are going to get involved in cigar play, you should purchase a good cigar ashtray. You can find a good selection at many different prices, just like cutters, at any good cigar store. I have a stainless steel one which supports a few cigars in the center of the ashtray. Should one fall off the support, it falls onto the stainless steel silver tray which supports the uprights not allowing the cigar to roll out. If you should go to the cigar store, ask the clerk about what they sell and what they recommend. Usually, cigar ashtrays are much bigger than a cigarette ashtray, and are deeper due to the larger amounts of ash and the greater lengths of the cigar. Again, they come in all price ranges and in all types of materials from plastic to crystal and should be at least eight inches or so to allow the majority of the cigar to lay in the ashtray. Cigarette ashtrays are smaller, generally 3-4 inches in diameter, and will not hold a cigar properly. It is very important to a cigar smoker to smoke their cigar in a safe environment if they are not going to hold the cigar all the time the cigar is glowing. A cigar, due to the glow of the ash being larger than cigarettes will set fire to anything it touches very easily. Cigar ashtrays also usually allow the ash to be protected at the ash end to avoid such mishaps.

WRAPPER, SHADES AND ROLLING

Although manufacturers have identified over 100 different shades of wrappers, only six are of great distinction.

<u>Double Claro</u> - Also known as "American Market Selection" (AMS) or "Candela", this is a green wrapper.

<u>Claro</u> – This is a very light tan color, almost beige in shade; usually from Connecticut.

<u>Colorado Claro</u> – A medium brown found on many cigars, this category covers many descriptions. The most popular are "Natural", or "English Market Selection" (EMS). Tobaccos in this shade are grown in many countries.

<u>Colorado</u> – This shade is instantly recognizable by the obvious reddish tint.

<u>Colorado Maduro</u> – Darker than Colorado Claro in shade, this color is often associated with African tobacco, such as wrappers from Cameroon, or with Havana Seed tobacco grown in Honduras.

<u>Maduro</u> – Very dark brown or black; this category also includes the deep black "Oscuro" shade. Tobacco for Maduro wrappers is grown in Connecticut, Mexico, Nicaragua, and Brazil.

To create the filler of a cigar, two to four different types of tobaccos are used. The filler is then rolled in a flat somewhat elastic leaf of tobacco know as the binder. Once rolled into a bunch, the tobacco is generally put into a wooden mold and pressed into shape for about one hour. When that is complete, the roller will wrap the molded and pressed cigar in a wrapper leaf which is supple, very elastic and visibly pleasing. The cigar is then capped and trimmed to uniform size. The finished cigar is then aged at least for 21 days. Many factories age the finished cigars for up to six months to let the different tobaccos marry and blend together for a finer taste.

There are a variety of sizes and lengths of cigars as well as various aromas. You need to go to a good cigar store, smell them, taste them, buy one of the brands to smoke, see how you like that brand and see how the cigar burns. You will want to have a good packed cigar that will maintain the glowing ashes as well as the smoked ashes until you get about a one and one half inches of burnt ash at the end before it threatens to fall off the cigar. This will give you endless opportunities of enjoying the cigar and playing well with it.

Now that you know almost everything about the cigar, here are some famous cigars and their barrel and ring sizes.

<u>STRAIGHTS:</u>

Churchill 7 x 48

Corona 5 ½ - 6 x 42 – 44

Double Corona 7 ½ - 8 x 49 – 52

Lonsdale 6 ¼ - 7 x 42 – 44

Panatela 5 – 6 x 38

Petite Corona 4 ½ x 40 – 42

Robusto 4 ½ - 5 ½ x 50

Toro/Corona Gorda 5 – 6 x 46 – 50

<u>FIGURADOS:</u>

Culebra- 3 twisted together panatelas

Diademas- 8 x 40/52 – 54 (closed and tapered head)

Pyramid- 6 – 7 x 40/52 – 54 (sharply tapered head and larger foot)

Torpedo- 6 – 7 x 40/52 – 54 (closed foot, a pointed head, and a bulge in the middle)

Belicoso- 5 – 5 ½ x 50 (tapered head)

Perfecto- 4 ½ - 9 x 38 – 48 (closed foot, a round head, and a bulge in the middle)

LIGHTING THE CIGAR

Let's start by holding the cigar in your fingers, rolling the cigar to see if it's still moist and doesn't crackle, or shows signs of aging or drying out. Next smell the cigar. I usually start at the open end and just run it under my nose until I come to the cut cap/head end smelling the tobacco and its aroma. The aroma should remain pretty much the same in intensity. If it doesn't, it generally means the cigar is old or dried out. You should roll the end of the cigar in your mouth wetting the end after cutting but before lighting. This will bind the end and give you a good opening for the inhalation of smoke. These techniques will give you a good basic taste of the cigar.

After you have chosen the cigar to smoke and have tasted and sealed the end, it is now time to light the cigar. You can choose to light the cigar yourself but more effective lighting is usually done by a second person, who I shall call the boy or submissive, but it can be anyone. Now comes truly a grand moment. The lighting of your cigar. To prevent a difficult or tight draw, or an unraveling of the wrapper, be sure not to cut too shallow or too deep. Using a cutter or punch, snip or punch the cigar at the shoulder of the cigar. The shoulder is where the head begins to slope toward the body of the cigar. If you prefer a punch-style, simply poke the sharp end into the head of the cigar, gently twist and remove. Although both of these moves sound like something erotic, these simple techniques will enhance your enjoyment of fine smoking..

Once the intense flame has been ignited on the lighter, bring the lighter within a short distance of the open end or foot of the cigar. This will "Toast" the cigar and prime it for lighting. While it is still warm, place the cigar in your mouth, or that of the other person, and hold it at a 45° angle over the flame. Slowly puff and rotate the cigar while maintaining slight contact with the flame. A good cigar will light easy and burn evenly. Don't let the flame touch the foot of the cigar. The cut end of the cigar, the head, should be in your mouth. Breathe in causing the air and the flame to ignite the tobacco leaves, and begin to rotate the cigar in your mouth to ensure the even lighting of all the leaves. This is critical to get an even burn. Should any embers appear, blow through the cigar to remove any unwanted odors that the flame may have caused such as butane, lighter fluid and or sulfur. Should you not do this, you can have loose embers and bad tastes and unwanted odors from the flame. You can rotate the cigar after the one side is lit and puff in to cause the leaves to glow which should slowly ignite the remainder of the leaves. You may even have to rotate the cigar a bit more to get a good even all over burn at the end of the cigar. This is important to ensure a good even removal of the ash. When you have to dump ash, it should be easily done with just a tap of the cigar. Be sure you know your cigar and how it taps and dumps before you begin playing.

You can also, after tapping the bulk of the ash, roll the cigar to get a pointed end if you're going to do precise burning. By rolling the cigar in an ashtray or on some other object, you will get a point at the end of the cigar. As you smoke the cigar, you have to understand what type of effect you want to play with and develop the ash to produce the outcome you want. For example: If I am sitting back, relaxing and smoking a cigar and a boy is servicing my cock. I may feel the need to drop some ashes down his back or through his buttocks, I will tap the cigar over his back or buttocks holding the cigar at least two feet from the body and allow it to

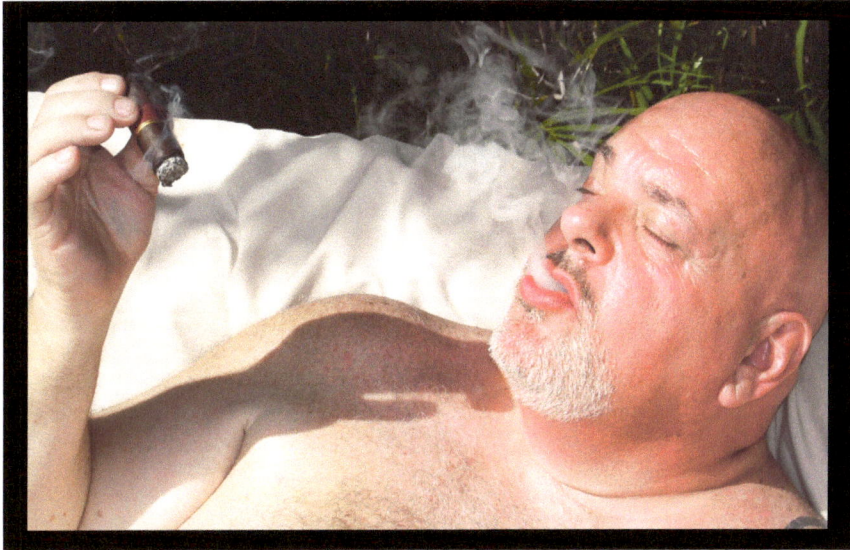

drop. I watch the ash to be sure there is no red burning cherry ash when it falls. A good cigar should never break at the red ash point. It should break just beyond the red ash part where the gray warm ashes are, which causes a hot warming sensation on your bottom. Depending on the intensity of that ash you can then judge how close to the body you want to play. Much like in wax play, the intensity of the ash is controlled by the height of the ash fall from the body. If it's higher, the ash has a chance to cool a bit and then the bottom feels less heat; if it's closer, the ash will be hotter when it touches the bottom due to the shortness of the fall time. Remember: don't butt out a cigar on the bottom as it will cause a third degree burn. A cigar will extinguish itself in a matter of minutes if not drawn upon and this is the proper way to extinguish a cigar.

I enjoy a good thick cigar with a nice smooth aroma. I like them well packed so that the smoke flows slowly through the cigar and allows a good build up of ash at the end. Smoke must flow freely through the cigar. If you have to suck hard to get the smoke, it doesn't burn well, nor does it smoke well, discard the cigar and start all over. My all time favorite was to see someone smoking a cigar that has a ring size of 64 gauge (almost 1 1/2inches in diameter). It was hot to look at, but not that great for play. I prefer a smaller cigar for actual playing. I always like to find a cigar which is firm and tightly wrapped so it burns slower and the ash tends to stay intack better, which allows for more precise cigar play.

Many leather boys/bottoms like seeing a massive glowing cigar in a Master's, Sir's, Top's mouth. It represents more ash and more possible opportunities of playing with hot ash as well as bigger burn areas. These types of cigars are great for singeing and we have a chapter on that a little later on. A smaller cigar is a more precise cigar and is good for small places, more intense ash marks and more accurate play. Like anything else in life, bigger is not always better. You have to just determine what the intent of the play is or the outcome of the scene and pick your cigar accordingly. There is also nothing wrong with smoking multiple cigars with different cigar thicknesses for different areas, or types of play. This allows the top to have a slow burning ash in one cigar, a big cigar for a large burn area and quick dumping of the ashes, and even other cigars for additional play or use.

The last subject I want to talk about is cigar smoke. Obviously there is a smell to cigar smoke. You should always do cigar play where either cigar smoke is allowed or outdoors where people can play with fresh air. Cigar smoke penetrates any fabrics and carpeting. If you're going to do this indoors please know that even long after the scene has ended, the smell will remain. Generally, deodorizers will not take out the smell either. I have even tried to steam clean the smell out of furniture and it doesn't do well. Once a room has been used for cigar smoke it generally never leaves the room unless you paint the walls and redecorate with furniture that hasn't been exposed to smoke. Great places to play are in a private patio, backyard or in a public space that allows both play and cigar smoke. Many of the gay bars allow this on patio areas or special vented rooms, as smoking has pretty much been banned from public areas in many states

We are almost ready to begin to talk about cigar play but let's recap just to be sure we all understand that size matters for the type and location of play. The moisture and wrap of the cigar also matters, for the burn of the cigar and the ash maintenance. There are some basic instruments needed for cigar play, a cutter, a cigar ashtray, a humidor, an assortment of cigars of your choice and a good cigar lighter for basics. Remember, that if you're going to do a scene with a cigar you can use one or many cigars, so like any scene, be prepared for the length and intensity of the scene which is negotiated with your bottom prior to play. Each of us will play at different levels and we must understand that negotiations are still part of this scene. Remember, you are playing with a hot, live burning instrument which can cause third degree burns. You can cause permanent scarring to someone with a cigar and this needs to be understood by all parties. If you play properly, play can be very safe, non burning, and very arousing with no marks or burns after the scene. Of course you can negotiate for burns and marks as some people enjoy being burnt and having "badge of courage" marks.

Before giving first aid, evaluate how extensively burned the person is and try to determine the depth of the most serious part of the burn. Then treat the entire burn accordingly. If in doubt, treat it as a severe burn.

- **First-degree** burns affect only the outer layer of the skin. They cause pain, redness, and swelling.

- **Second-degree** (partial thickness) burns affect both the outer and underlying layer of skin. They cause pain, redness, swelling, and blistering.

- **Third-degree** (full thickness) burns extend into deeper tissues. They cause white or blackened, charred skin that may be numb.

  Remember, your first aid kit and have it handy, just in case an accident happens. Also, be prepared to take your bottom to the emergency room or call 911 should you have an accident. Burns are not something you should play with, especially when skin has been removed and raw flesh is exposed. Another good item to have around for safety is a wet towel to immediately put on a small burn to help cool and protect the skin. By giving immediate first aid before professional medical help arrives, you can help lessen the severity of the burn. Prompt medical attention to serious burns can help prevent scarring, and deformity. Burns on the face, hands, feet, and genitals can be particularly serious.

- **First-degree burns** usually heal in 3 to 6 days.

- **Second-degree burns** usually heal in 2 to 3 weeks.

- **Third-degree burns** usually take a very long time to heal.

This type of play requires a good deal of trust between the two parties as well. Trust is one of the key words in BDSM play and this type of play certainly qualifies for that word to be understood.

How do you find people who are into cigar play? Well, the color of the hankie is tan. Obviously worn in the left pocket, means a cigar smoking top. Right pocket, means a cigar smoking bottom. You can obviously see a cigar smoker or smell a cigar from a room or so away and sniff down your top, but that doesn't mean they are knowledgeable about the play. I always tell all bottoms in any class I teach, be sure to quiz your top on how they play or where did they get their education in BDSM. Questioning them about their play and seeing their responses before you make your decision to submit is critical to help you make your decision on whether you want to extend them your trust. Only you know when you feel safe and worthy of submitting to that top. If your gut tells you NO, remember, your gut generally is right and don't play with that top. Usually, your gut knows more than your hard cock or aroused body from the smell or sight of a cigar smoking top!

HEAT VS SMOKE VS HOT ASH

There are three basic types of cigar play which will have some techniques discusses within each major category. We will discuss these techniques under the major subject of each one. **Heat, Smoke and Hot Ash.**

HEAT PLAY - which is done by holding a cigar close to the skin or body part and then inhaling on the cigar to cause the cigar to glow. Remember, this is a hot item you have in your hand and only with experience can you determine how close to go to the skin or body part to get a reaction from you bottom. You can heighten the sensation by blindfolding your bottom, or maybe you are after the fear factor by letting the bottom see the glowing cigar coming at him. Both are effective and you will find this can be a very hot way to play. I recommend that heat be used only after you have your bottom warmed up. This is the most dangerous part of the play, which is why I address it first. Some bottoms actually like to be burnt with a hot cigar. I do not recommend burning your bottom for any reason. This is not safe play nor healthy play, especially if someone is diabetic, or HIV positive or has any type of skin disorder. Remember, that if you burn your bottom not only does the bottom have to scab and allow the skin to repair, therefore making your bottom not available for play, but it can cause serious and long term effects on the skin such as scarring. Third degree burns are very common with this play and again I stress, I do not recommend burning your bottom. Usually hospital care is necessary with a third degree or higher burn. A little later on I will discuss where a hot ash can be put down on the skin and how you can prepare it. A lot of people confuse burning the skin with this hot ash play. There are also some very pain orientated bottoms who enjoy burning, which we will discuss later in the book also, but again I don't find it necessary. Play can be just as effective with heat as to burn your bottom but I will address it later on and let you make your own decisions and negotiations. Also there are bottoms who like the "badges of courage" known as scars on their bodies from their play sessions, but all this should be done in the negotiations.

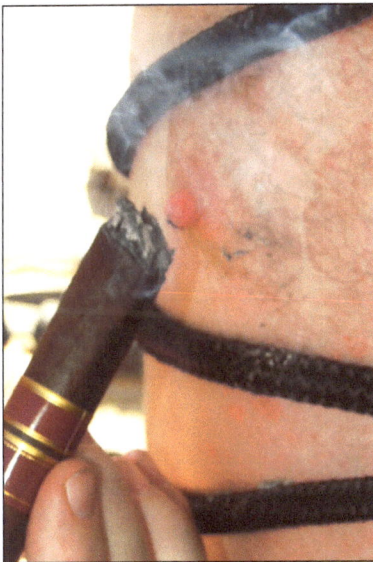

Heat is very effective on a nipple area, the opening of the ass or sphincter muscle of the ass, a ball sack, a cock head or underneath the entire length of the cock, and last but not least, arm pits. When I know the bottom and know he doesn't fear the hot glowing part of the cigar, I will have the boy hold his tongue flat in his mouth and open his mouth warning him not to move and insert the cigar into the opening of the mouth and slowly puff on the cigar causing heat in his mouth. The top has to be careful here again as this type of play can cause a burn on the roof of his mouth, if the bottom panics, which is very painful.

Heat is also used in singeing.

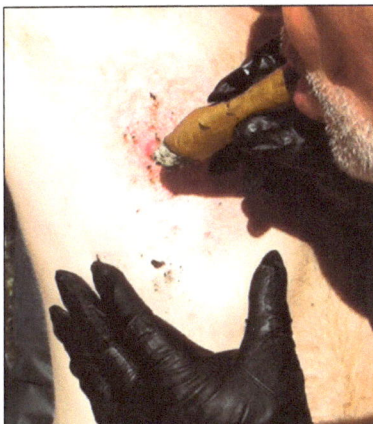

SINGEING

Singeing is where the glow of the cigar is used to burn the body hair. **Do Not** use on the face at any time! Hair is highly flammable, so when doing this type of play remember fire burns upward. Start singeing at the top and burn in layers down. Singe only until you are close to the skin never all the way down. Think of a mountain of trees. If the forest fire starts at the top it usually doesn't burn down much as fire climbs. If the fire started at the bottom of the mountain, the trees will all burn upward as fire flows upward catching the next line of trees on fire and so on until it reached the top. If you should start the hair on fire starting at the bottom, you can burn the whole area in a fast uncontrolled burn. Remember this is a controlled burning by singeing, not uncontrolled. The top must always remain in control and ready to douse out any fire should it occur. Singeing usually melts the hair in a way, instead of causing a flame, but too much singeing at once could cause a fire so do it slowly and methodically. Therefore, like in wax play or fire play, have safety items such as wet towels, water and ice close by, within arm's reach, for emergency needs.

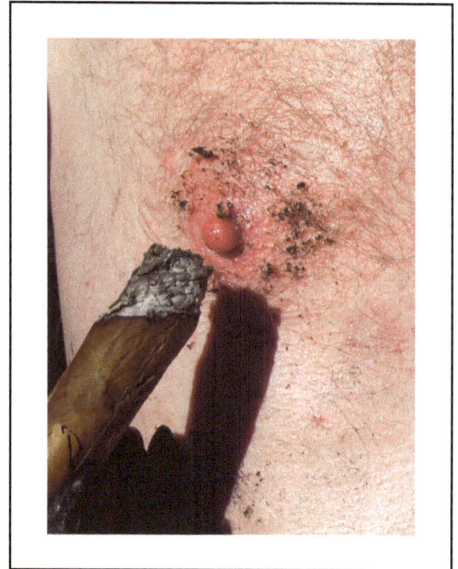

I love to singe hair in the pubic areas taking the boy's pubic area down and then shaving him. I also love to singe through the ass of a boy, around his nipples and his arm pits. Hair does have a distinct odor when it burns so be aware of it and understand that this can also cause fear or be an erotic turn on for some bottoms. Remember, singed hair does not taste great. If you are going to lick, kiss or suck any part of the body where you have singed the hair, be prepared to taste the singed hair.

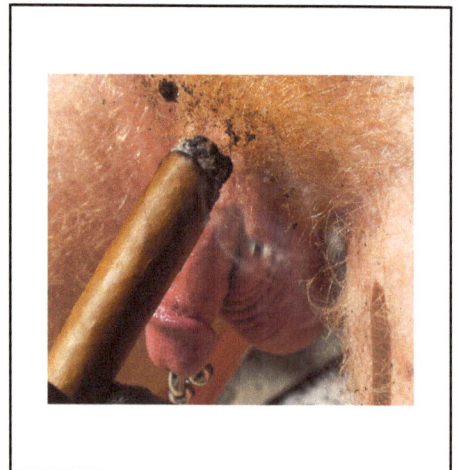

If your bottom is lying flat on a bondage bed or in some position other than standing, make sure you take the basic fire understanding of burning upward under consideration. I generally start singeing at the base of the cock, then moving to the balls, working down the balls, when the boy is laying down working my way down to his crack. The same procedure would happen if the boy was standing up, as his cock would be higher than his balls. You can also start at the nipple and singe down his chest, as usually the nipples are higher in the body than a flat stomach. Remember though, on heavy set people this could be slightly different due to the body curvatures, so judge the body and remember the fire rules of burning upward.

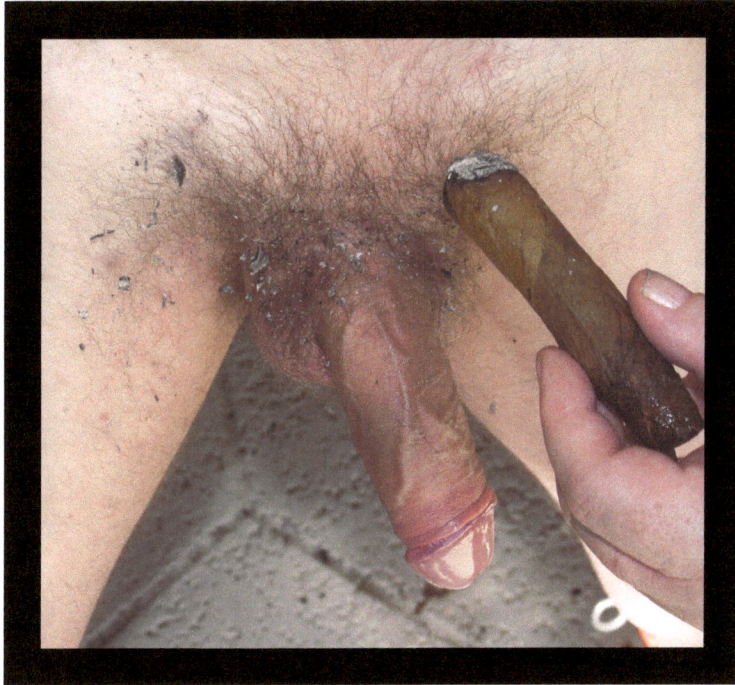

Singeing is very erotic to me and I force the bottom to stay still, developing a very good submissive control to avoid burning him. The fear of being burnt is a great control factor in this play. A bottom who submits to me in this manner, excites me greatly, as he truly understand he is surrendering his body. The smell of his burning hair excites me and may scare or frighten him as well, this means he truly understands I'm using my power to take his hair from his body through his submission and control. He smells his hair burning which causes him to believe he is surrendering everything to me. This smell allows my power over him to take whatever I want from him. This same smell of burning hair makes him feel he is totally not in control. I on the other hand have to ensure he doesn't get burned from moving his body. I recommend using bondage to help control the submissive's body. Trust here is a very important factor as hot glowing ash is necessary to singe the hair as well as a heat factor is present to the bottom as he smells the burning hair from his body.

SMOKE PLAY

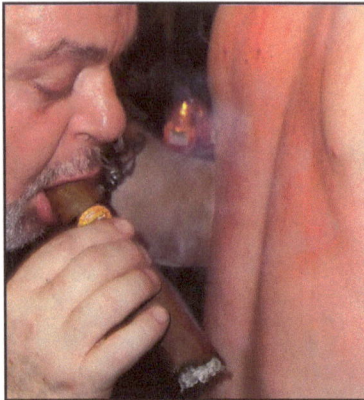

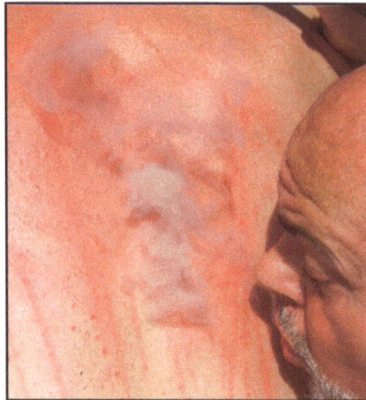

Yes smoke play is very erotic play. I use smoke as the starter to my cigar play. It allows me to see how the boy will handle the smoke. I then move on to the heat and the ash seeing how he will react before I begin to do any type of play for extended periods of cigar play. Remember, the sky is the limit with smoke play. Smoke can be created to take shapes and used in different ways in kissing and other types of play such as ring toss with the boy's cock. Smoke is relatively safe play with no chance of burning. The only real danger in smoke play is that taking in smoke can cause shortage of breath and or choking to your bottom. You just need to be prepared for that reaction. You can use smoke in kissing, from a simple kiss, a French kiss and finally to a Masters Kiss. What is a Masters kiss you ask? Well that is where a boy is kissed with a mouthful of smoke passing smoke to him and back to you in a rhythmic breathing taking kiss, giving the breath and the smoke back and forth. The Top or Master will hold the boy's nose tightly closed as to not allow for any air or leakage of smoke. The top can control this kiss until the boy almost passes out. This is a sort of breath control where you are breathing on the left over oxygen from the breath passed back and forth. If you have a good winded bottom that can last a long time, you can inhale additional oxygen through your nose adding more oxygen to the breath. Hold your bottoms nose tightly while you take in the additional oxygen through your nose. You then can extend the kiss process much longer this way. Your bottom will have little new air from this intake of air and you as the top, will feel somewhat refreshed. The addition of smoke enhanced deprivation of oxygen can help further the bottoms submission, again having a power exchange through this means of play.

Smoke can be blown onto any body part such as around the cock and balls or rimming an ass as well as blowing smoke up the submissive's hole. You can literally blow smoke into a sub's face as a torture. If you are talented with your smoke, you can blow smoke rings and have the boy's cock catch the smoke rings for some real fun type play.

Smoke has a warm sensation and is relatively safe in play but the top has to be in control of his cigar during all this time, as the boy will be moving. Some people however, really don't like smoke, so be sure you cover this in your negotiation. Generally smoke loving bottoms or cigar bottoms love smoke play as the smell and the taste and the feel of the hot smoke on and around their body excites them to submit and do more play. You can also give a good oral session with smoke which we will discuss further as a special chapter on oral play and servicing your top.

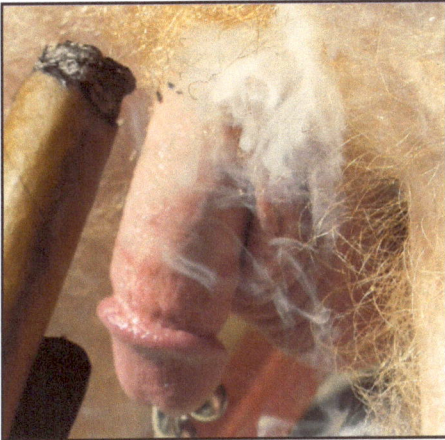

Remember, in any type of cigar play, the top must remain in control of the cigar. If you hold your boy in any way, shape or form, always try to keep the hot glowing ash end away from him using your hand as a barrier from the hot end of the cigar. A boy can always move and you want to take as much precaution and protection of your bottom in play with any hot item. If his hands are behind his back and tied, for example, remember the cigar may drop hot ash into his hand so be aware of the surrounding and possible circumstances that can happen with the falling ash and possible embers. Protect your boy as much as possible. Put the cigar behind your back when his hands are tied so as to protect him from any burning. But remember yourself and the same applies to you; keep the hot ash away from your body by positioning your hand appropriately.

HOT ASH PLAY

Hot ash play is when you use the ash of the cigar which is not glowing. This is why in the beginning I stated you wanted a cigar that holds about a one to one and one half inch of cigar ash after the glow. This ash works much like wax play. It is still hot to the touch but not burning. Hot ash cools relatively quickly. The red burning embers are known as the cherry of the ash, and should never be used in ash play.

If you are going to write with hot ash on a body be sure you have swirled your ash in an ashtray to form more of a point to the ash. This will allow you to write on the boy's body and have more control of the ash, as it is a smaller point much like that of a pencil. I love to put a boy in bondage and then do cigar ash play, writing "slut", "whore", "insert here", "pig", etc. on the boy's body. There are some dangers here as the ash can break off, so be sure your ash is attached well to the cigar, and work lightly and quickly. If you see the ash crack, stop and dump the ash. Only use the non-glowing ash of the cigar to do any hot ash play. Never use the cherry ash for writing as it will cause third degree burns.

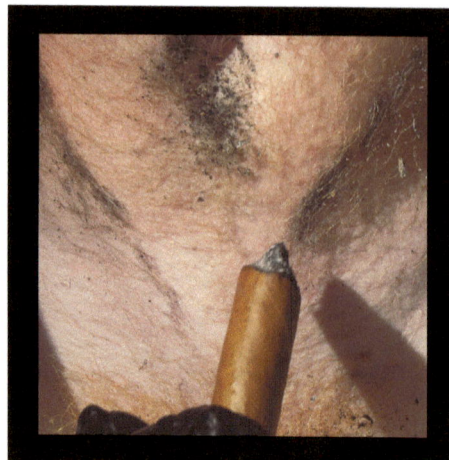

NIPPLE PLAY

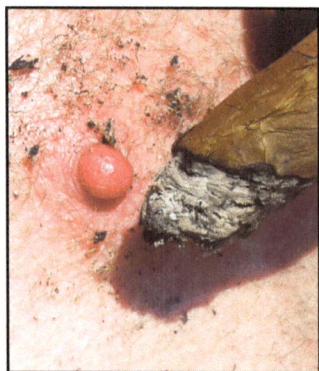

Nipple play, I recommend using an ice cube to get the nipple aroused. It will also allow the skin of the nipple to tightened up. Then slightly push the cigar straight on into the nipple, using only the non glowing ash. I usually sharpen the end somewhat so it doesn't fall all over the nipple and skin, causing it to break off. The residue of the ice causes the nipple to be slightly wet and helps in cooling the ash. This will leave a nice ash mark and the boy will go wild with his nipples being used this way. Again, practice this procedure many times before touching a real nipple. Start slowly, go light handedly on to the nipple, watch for the pull away. NEVER smash the cigar onto the nipple. Trust me, with just the hot ash on the nipple you will see great results in the bottom and his reactions. If someone is really into nipple play, this can go on for a long time repeating this process doing several applications. Just remember, the nipple will crust over much easier than many other parts of the body, and will burn less, as the tissue here is denser. The coolness of the ice and the hotness of the ash will cause an increased sensation for the bottom with less hot ash making this process much safer. Again be aware of the cherry ash on the cigar and how deep the hot ash is before starting. After the last puff of a cigar, allow the cigar to cool down a bit and allow the cherry ash to cool or turn to hot ash before using it.

COCK AND BALL TORTURE

Heat and ash play can be done around and on the cock and ball area. The same practices can be used with any part of this area. Just remember to go slowly and lightly. A little hot ash goes a long way in this very sensitive male area. The head of the cock however, is much more sensitive so proceed lightly. Ash play like this basically can be done anywhere on the body once you understand that you are using the non-glowing part of the ash. I again repeat, NEVER USE THE GLOWING CHERRY ASH.

Smoke surrounding the cock will cause the cock to feel warm and help increase circulation of the cock. Again ball singeing and ash play can be done on the cock and ball area using all the principles taught in regular play with these two types of play. Engulfing the testicles in a warm smoke filled mouth can have a multiple different type of feelings depending on the bottom. Ice can also be use to cause a great difference here in temperatures both in heat and in ash play. Using bondage on the cock and ball area also increases the sensitivity of this area which will heighten the feelings of the smoke and ash. Really any form of cock and ball play adding smoke and ash can be done. A simple spanking of the balls then blowing smoke on them or ash droppings is extremely great. Electrical ball torture then adding smoke or writing on the ball sack with a cigar can be excruciating fun. Once again, the sky is the limit; just add smoke, ash, or burning to any existing play and you will find a new kind of play.

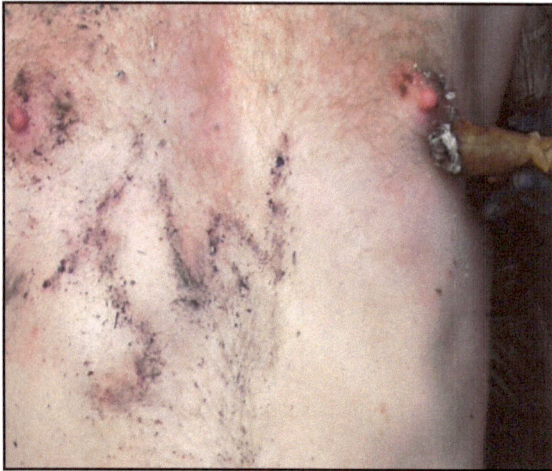

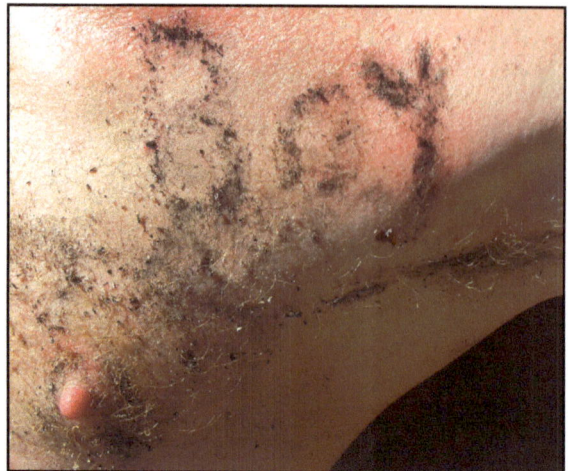

HOT ASH WRITING

You can do designs, drawings, and the your imagination is the only limit with what you can do with ash. Remember, to keep smoking the cigar as you use the ash allowing the ash to cool before you use it. Should the ash fall off and break at the red cherry glowing ash just puff on the cigar a few times and allow the ash to build and cool before using it on any part of the body. You may have to re-sharpen the point or make a new point to the cigar before you resume writing. The ash will remain warm but not warm enough to cause serious burning. Cooling the body with an ice cube will also cut the burning sensation and protect the skin somewhat. Remember, the body part where you are ashing, and ask yourself what the skin is like there. Is it thinner skin or tough skin?

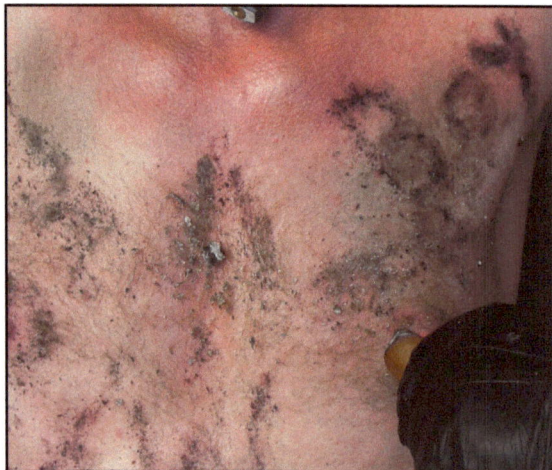

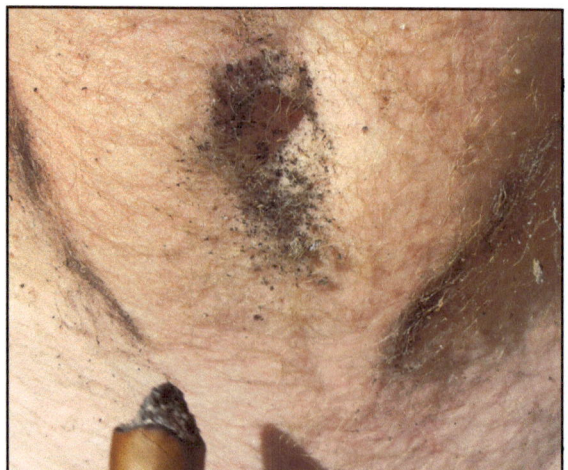

Approach the skin accordingly? Ice can cool the skin making it tight and wet. This wetness will always slow any burn down. I have even gotten some great sounds from this technique when the hot meets the cold. A slight sizzle sound can occur from the wetness. It causes more extreme feelings but it doesn't burn nearly as intensely, being that the skin is wet and cold. Remember different size cigars here in different areas cause different types of writing and cigar ashing. Rolling the cigar ash will make a point to help with writing and cause less breakage.

Just remember, if you are using ice you are going from a cold temperature to a hot ash rather than from the normal body temperature to a hot ash temperature. Therefore, your bottom will receive a greater sensation of degrees difference. The bottom feels like a bigger difference has occurred, when really you have taken the skin to the cold side or chilled side first and then warmed it back up. The bottom feels he has accomplished a bigger degree of intensity difference, making him feel more submissive and you are really more in control of the burn capabilities. Just remember it can cause additional damage of too much cooling to a hot cherry type of burn. Cooling is only good when the skin really isn't broken.

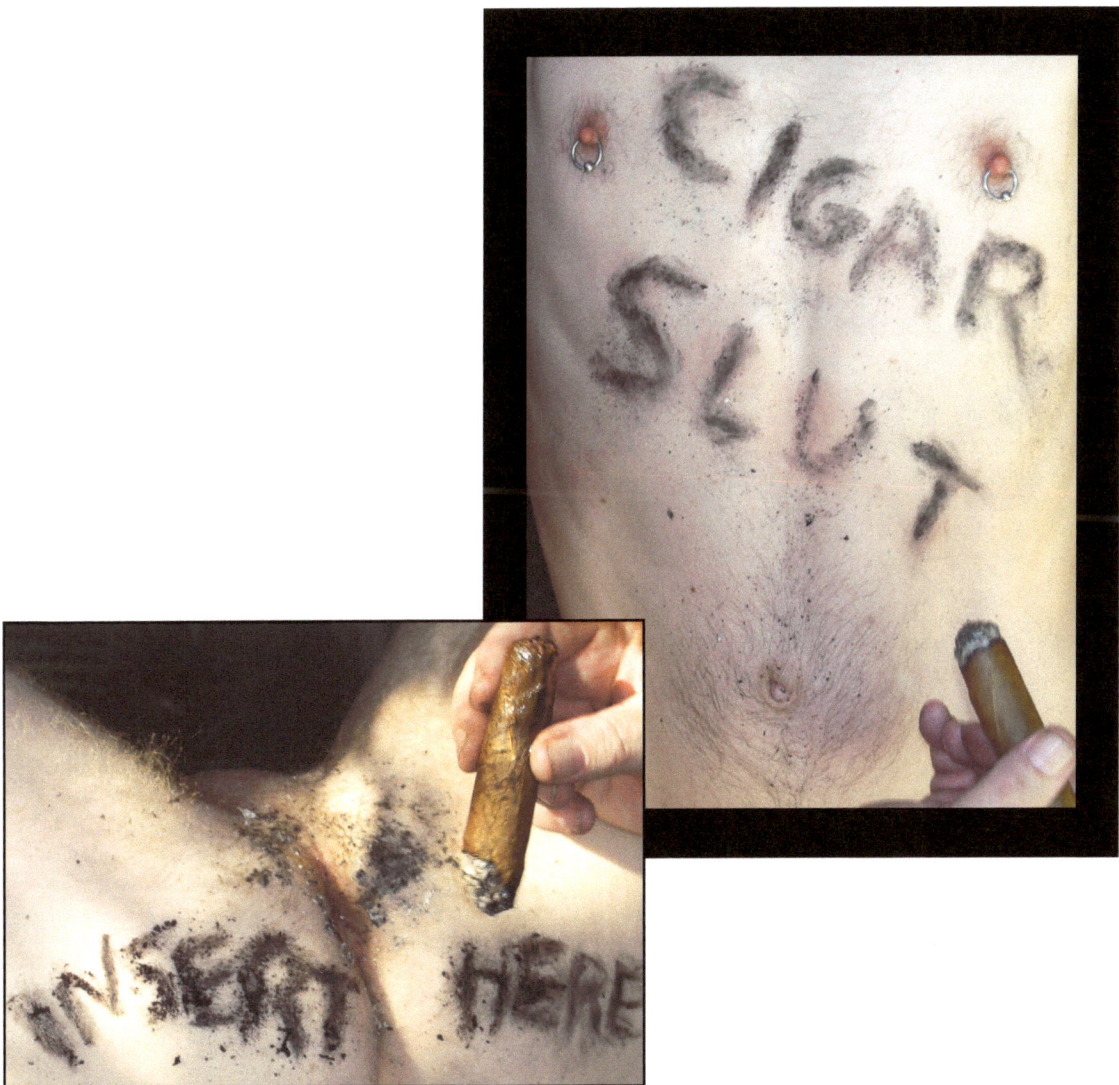

KISSING WITH SMOKE

Kissing with smoke is a very erotic type of play. The top usually controls the smoke and I say usually, because the intake of smoke can be done by either partner. When whichever partner intakes the smoke and begins kissing the other partner with smoke, he allows the smoke to transfer into the other person's mouth either forcefully or passionately.

You may allow some smoke to exit the mouth going up into the nostril. You can be forceful and blow hard, forcing the smoke into the kiss not allowing it to escape or you can also be passionate with smoke and do a lot of tongue with smoke exiting and licking and spreading the smoke all over the other's mouth or body part.

The nipple is another area of eroticism that works well with smoke and warmth in a passionate kiss. You can lick and touch and kiss and allow the smoke to caress the nipple arousing it with such warmth from the saliva and smoke. Again, it can be a simple kiss. When I kiss with smoke, I tend to allow the submissive to view the smoke coming from my mouth, let him smell the cigar I have chosen and many subs generally get sexually excited even from this simple vision.

I once asked a submissive in one of my classes why he enjoyed cigar play and he said it reminded him of his grandfather, whom he loved deeply. The grandfather used to smoke his cigar when he was little boy and would always give him a light kiss. The cigar and smoke reminded him of that person and time in his life when he kissed him. He simply was melted by the cigar smoke and the smell in remembrance of his grandfather, as the dominant man he was, controlling the family and giving him that special grandfather kiss.

EAR WAX REMOVAL WITH SMOKE

I read an old family medical book and I took this old medical treatment and turned it into a type of smoke play. You can actually do wax removal from the ear in smoke play and clean out the boy's ear. You also do something very good for anyone's ear by blowing smoke slowly into the ear. This will help melt the ear wax. The warm smoke has medicinal benefits and allows the wax to become soft and melt. You can do a whole scene like a medical cleaning of the ear wax by blowing smoke and kissing the ear. You can add control to this scene by putting the bottom's head either in bondage or some other form of head restraint. You will need to form a cone with paper or some other type of soft material, allowing deeper penetration into the ear. You will blow the smoke into the cone which allows the smoke to fit into the ear canal. The cone delivers the smoke more concentrated. Do this a few times to ensure the smoke penetration and that the warm smoke starts melting the wax. Just remember you need to remove the wax once it is loosened by the smoke. I later found out that this was an Indian way of removing wax from an ear. It can be very sexually stimulating feeling the slow internal hotness of smoke into the boy's ear. You can even help a sore ear with this type of play, so you see, it is a very safe and highly beneficial play for someone as long as you are educated in this play. Remember, never stick anything sharp into the ear and be sure to clean the ear after playing like this.

CIGAR PLAY BREATH CONTROL

First off, bottoms who enjoy breath control will generally like breath control with smoke. It adds a whole new dimension and more of a power play. Let's begin with telling you that you can use a gas mask, your mouth, or any other way you choose to control the breath and smoke to encompass it all for your bottom. Let's start with doing breath exchange, a form of mouth-to-mouth.

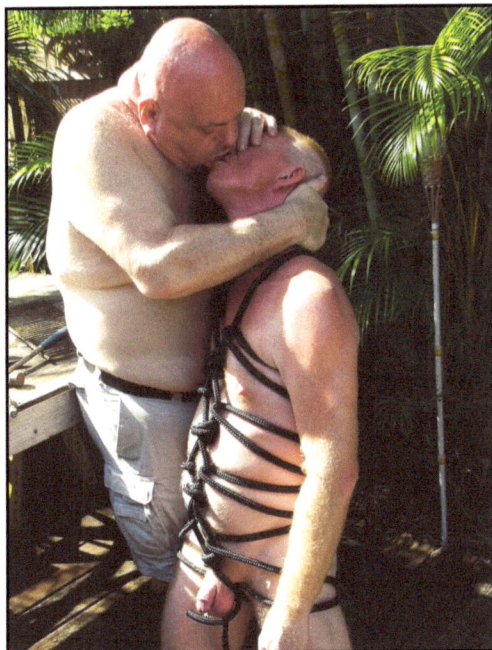

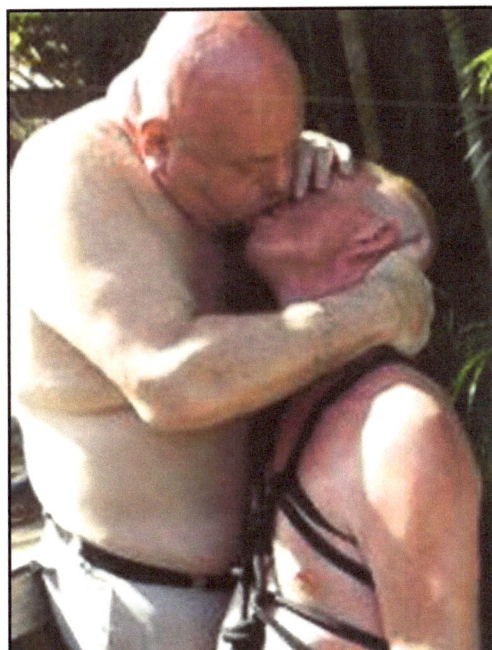

BREATH EXCHANGE

Begin with a breath control type of kiss, suppressing the nose of the bottom to seal his way of exhaling except through his mouth. Begin to synchronize the kissing and flow of air back and forth between the two of you. Now the top takes an intake of smoke from the cigar and sets the cigar in a proper ashtray, not allowing the cigar to roll and set something on fire. He then, forces the smokey air into the submissive's mouth and the submissive should take the air and smoke in. When the bottom has had enough of the intake from the top, the top needs to pause for a moment then take back the air, sucking it all the way out of his bottom. Then when he feels he has taken all of it back from the bottom, the top then returns the smoke air into the bottom's mouth forcing it down back into his lungs, without taking any more oxygen or smoke.

This is known as a Master's kiss. Two things happen here. The bottom is living on a dwindling supply of oxygen as with each breath, more and more CO_2 (Carbon Dioxide) enters the bottom's mouth, making it harder and harder to breath. He also gets the smoky taste of the air. Once again, the bottom passes back the air to the top and this can go on for a number of times, back and forth, each time increasing the CO_2 and causing the bottom to become light headed or dizzy from the kiss of smoke. The bottom can also just release the smoke which takes the Master's kiss to just a passionate kiss, letting the bottom enjoy the smoke and the breath exchange from his Master.

In a Master's kiss, if the top needs to add more oxygen, he or she can take it in through their nostrils keeping the bottom's nostrils and lips sealed, so that the scene can continue on for a number of passes. Be sure to be aware of your bottom's coherency and alertness, because as in normal breath control every bottom differs in their intake of oxygen and CO_2 differently daily. Repeated applications of this play or just a simple passionate kiss can not only be very passionate and erotic but can be very submissive for the bottom. They tend to enjoy the light-headed space and smoke taste from the cigar and shortness of oxygen and the passion of a hot kiss unlike that of just a regular kiss. The smoke adds another dimension to just a regular kiss and the ability to breathe. Breath control with smoke is an added feature adding warmth and many other attributes to a basic scene.

CIGAR PLAY WITH A GAS MASK

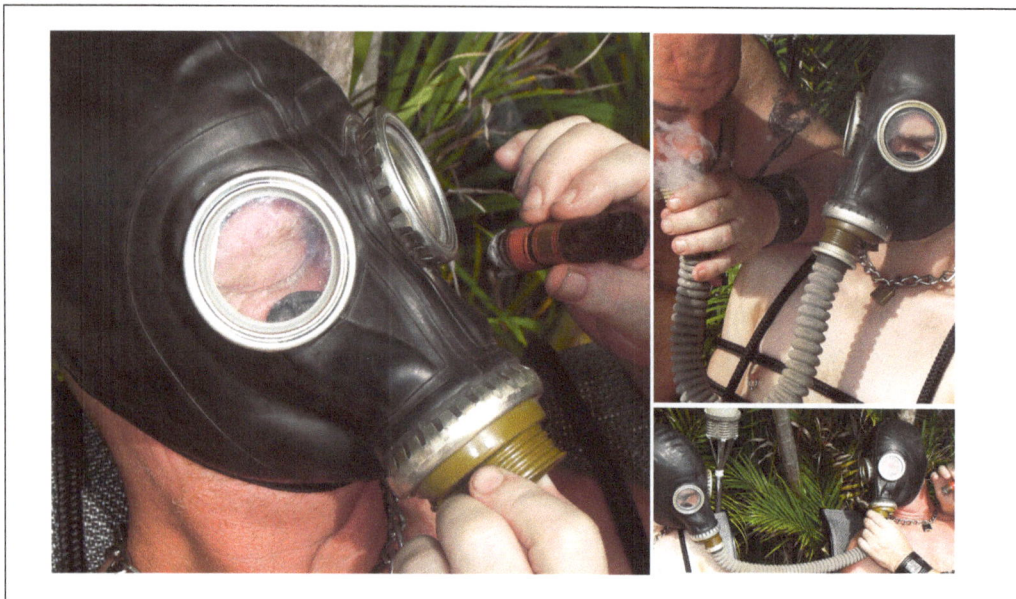

The bottom is fitted with a tight fitting gas mask which can be purchased from many sources. I usually like to buy them from the Army Navy surplus store to add authenticity to the gas mask and like to have the bottom in some sort of army fatigues, or gear, to add to the scene. I also like to add some bondage to this type of scene to help prevent the bottom from pulling off the gas mask. I make him do some calisthenics or some other type of exercise causing his body to get the blood flowing and rushing through the head. I then put the gas mask on to the submissives head ensuring a tight fit. I then begin to check their tolerance to the air restriction through the hose of the gas mask before I begin to add the smoke. If they are breathing fairly normally, I then inhale the smoke from the cigar and blow it slowly into the mask. Be sure to blow slowly, as the bottom is already taking his entire air intake into the mask through the hose. By him breathing through the hose only accelerates the intake of the smoke into their mask and lungs. Be sure to have the bottom close his eyes as smoke can cause tearing. Depending on the bottom, and the sadist or the top, tearing can be a good thing however. Watch for choking or other reactions. Then seal the end of the hose controlling the breathing and watch for the panic to set in. The bottom does not really have any access to clean oxygen at this point but what oxygen they have is smoke filled causing the panic and shortness of breath. This is the danger point so start off slowly until you learned your skill and watch for signs of passing out and or choking. This is where the educated top will shine.

Breath control has a high level of danger and panic, so know your bottom and know what they can handle in breath control before starting this type of play in a gas mask. The smoke reduces the amount of oxygen so they can pass out quicker, so be prepared.

With the slow reduction of oxygen and the smoke the bottom can go down either quickly or really slowly depending on their level of play. Should the bottom pass out, remove the entire gas mask as quickly as possible and be ready to administer CPR and breathe into the bottoms lungs clear, non-smoke air from your lungs. Below I show two bottoms using each other's air in deprivation with the use of two gas masks.

I enjoy this play but also know that this can cause a panic attack before the bottom runs out of air. You should have him restrained so as he does not try to pull off the mask or do other damage to himself. You must be in total control of this scene as the top and know that, with each breath the bottom exhales, their release fills the gas mask with more CO_2, making the air harder and harder to breath. Watch for the color of the bottom's skin as they will start to lose their normal coloring. Some bottoms will turn real red, others will begin to get a whitish greenish look to them, as the oxygen become less and less part of the intake and lowers the level of oxygen in the blood. Each bottom is different, but you need to be aware of the signs as well as their alertness.

CIGAR RIMMING

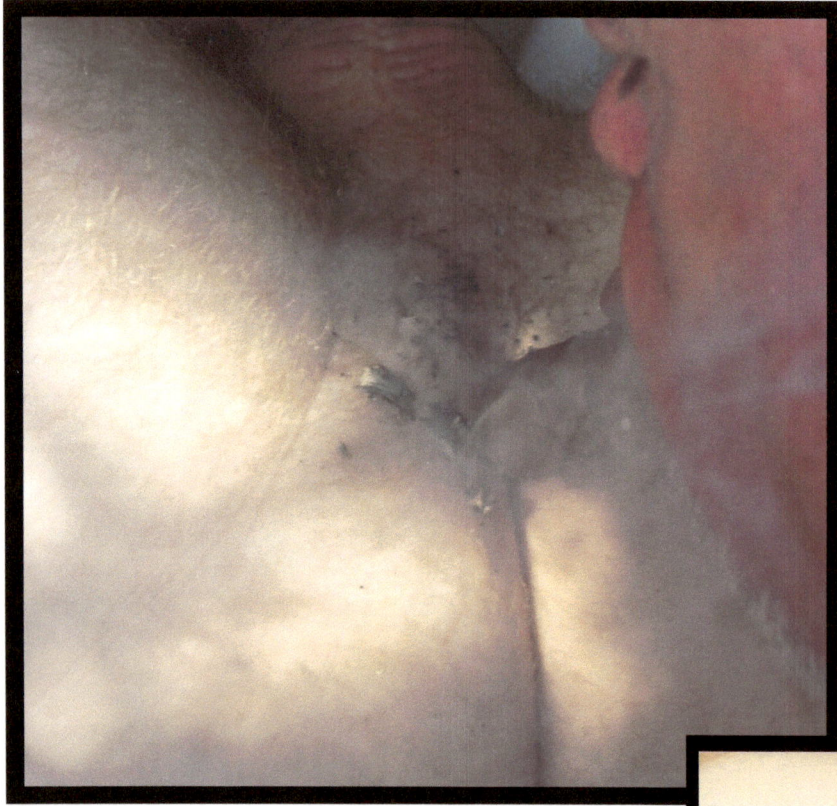

Cigar rimming is a quick and hot way to pleasure someone, causing a very different feeling from just a regular rimming. If you don't know what this is, it's the actual oral stimulation of the anal region either by tongue movement or by sucking or by using your facial hair. The other person still feels the rimming, but adding smoke to this type of play makes the tongue warmer. The smoke makes if feel hotter but you can blow smoke into the anus and the other person will feel the warm heat entering him causing him anal stimulation of a different sort. Rimming is a low risk play compared to popular belief, provided there is no blood or cum present in the rimming session. You can also use a cigar and stick the head end of the cigar into the ass region of the bottom taking his body flavors onto the cigar. Now we are not talking a Monica Lewinsky scene but really it is just a different body part scenting the cigar.

You can also train your bottom to inhale a bit. The cigar will take in some smoke and it will look like they are smoking a cigar from their ass. I find this very hot looking and lots of fun.

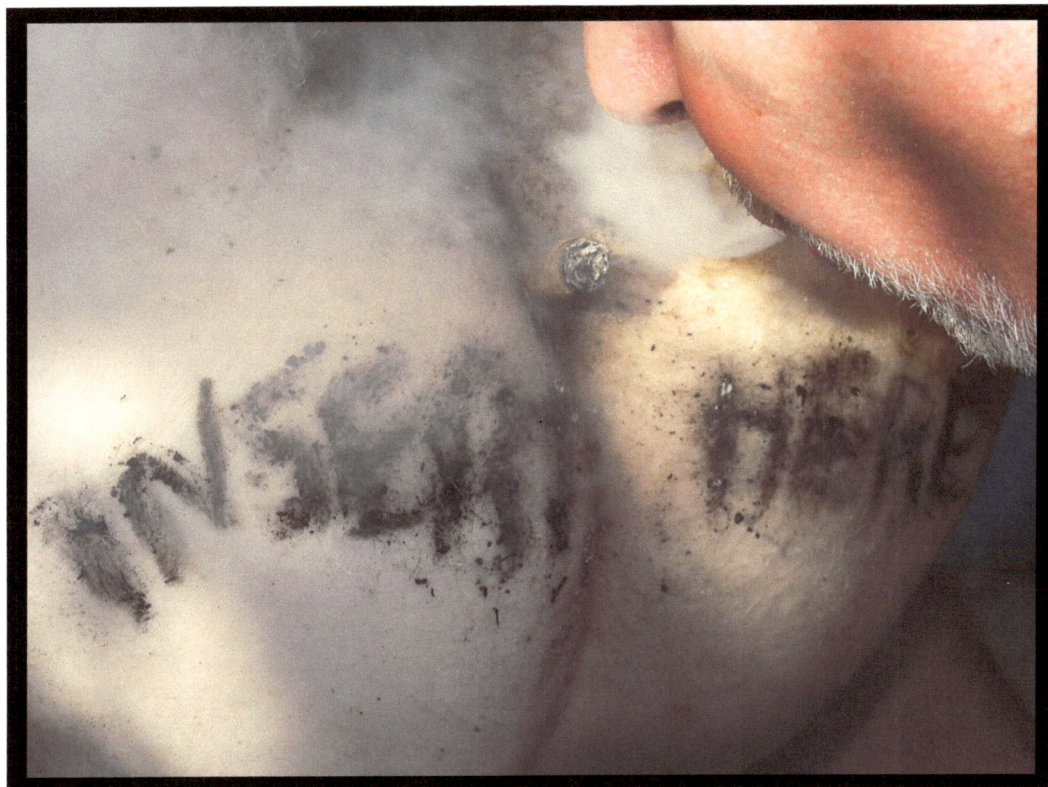

However, if you decide to insert a cigar into the bottom's anus, don't do it very much or for long periods of time. The anal walls will become very irritated by the nicotine in the tobacco. The walls of the rectum are very sensitive to the nicotine and the rapid absorption can cause a harsh sickness, especially for a non-smoker. (The same applies for the vaginal walls) Smoking it afterward is something for you to consider seriously; the lube on the cigar isn't going to make the cigar taste very good. Of course the old "spit-n-shove" technique is always fun and doesn't taste as nasty. Using a cigar like a dildo leaves the possibility that the cigar may break or some cigar leaves may flake off inside the rectum. Just rub the cigar between the cheeks after licking the ass opening and pick up the musky ass smell and sweat taste is my recommendations.

CIGARS AND SERVICING THE TOP

Well here is a subject I love to talk about, for I love having this done to me and playing this way. It is not only a very relaxing and comforting way to receive oral service from your bottom but can also be where your bottom and you can grow from the servicing. Obviously, and I hope it is obvious, the word servicing means taking care of the tops needs. I usually insist on being almost naked here, for I enjoy having my entire body licked and serviced by my boy and eventually to have oral service on my cock. I usually start by being naked and having him start at my feet so I can allow myself some self-indulgent time with my cigar. I relax on a bed or in an overstuffed chair, extended out, so he has access to my entire body. I puff on my cigar and blow smoke in his face as he begins to lick and service me. I eventually give him a kiss for a good servicing and use that time to enjoy his mouth and tongue while I enjoy my cigar. Eventually, if the boy smokes cigars, I will pass him the cigar and enjoy the servicing further with him blowing smoke across my body. Then when he is giving me head, with smoke in his mouth, the smoke will be pushed down into his lungs by my cock. Much like in the rimming section, he is allowed to enjoy the cigar and smoke, taking the smoke in across his tongue. He then uses his hot, wet smoky tongue across my body licking, kissing and feeling all the parts of my body sensually. Or he will take the smoke from me in a hot kiss, and then do the same, as if he was smoking the cigar. He will enjoy my armpits by smoking and then licking the hairy pits and the smell of my pits will cause him an enjoyable renewal of our bodies together. If it is after a long, hot, sweaty play, my armpits being hot, sweaty, moist and smelly with the addition of the cigar smoke running through his taste buds, plus all my smells, all become very hot. Around my cock the same will apply. The smell and taste of precum or after fucking the smell of his ass and my cum, will add to the flavors of my cock region. Imagine yourself relaxing and having your balls sucked with a hot, wet mouth tasting the mustiness of your manhood.

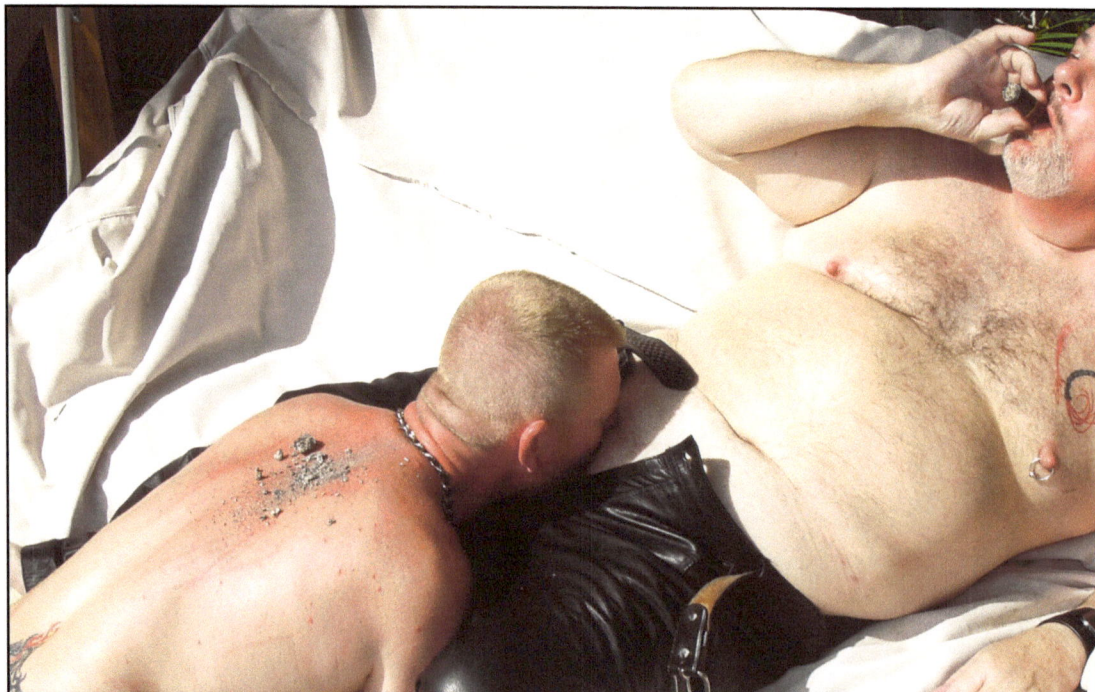

How hot is that! I enjoy being rimmed as well, so he does the same as I did to him in his ass, other than the insertion of the cigar in my hole. He may roll the cigar in my hairy hole to allow my flavors to envelope the cigar much like a Monica but it is not inserted into my anus. This is my request but all is open to you when you are making that decision. All of this oral pleasure with a hot, wet, sweaty, smelly, Master's smell cigar in his mouth, makes both him and me erect and ready to explode. Even writing about it gets me hot and excited so I hope it does you. Close your eyes and just imagine the feelings and the sensations and the tastes of his body or yours. Woof! All this and it is perfectly safe; oh my God what pleasure!

If you like to fuck, fuck his hole smoking a cigar, allowing the ash to drop on his ass and back, causing arousal. You feel him milking your cock into his ass, making you both filled with all sorts of arousal and pleasure. Wow! Now that is the ultimate relaxation to me… Woof…..!

CIGAR PLAY DOING BURNS

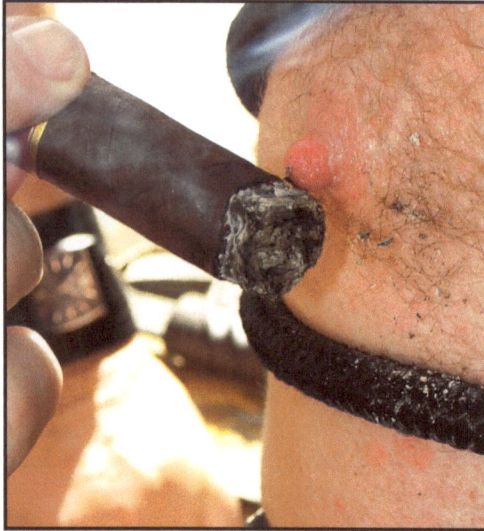

Cigar play actually burning the skin is something that only should be negotiated with very experienced bottoms. I know one super pain bottom who will take a red hot cherry ash and literally allow it to be pushed into his skin, burning the skin to a third degree burn. He likes the endorphin rush and doesn't mind the scarring, but overall this type of bottom is very hard to come by. I will discuss doing direct burns here using red hot cherry ash now.

When the cigar burns the skin, the temperature is greater than with normal body temperature. The bottom will feel a greater burning sensation, although the burn is the same process just deeper, causing more damage to the skin, he will get a much greater endorphin rush. I love doing this on a nipple where the skin is much tougher. The nipple can crust over much easier and replenish itself without much issue, compared to that of the flat epidermis level of skin. You also have to consider how hairy the area is you plan to burn. You will be singeing the hair on the way down and can cause a major burns as well as a possibility of the hair burning uncontrolled.

Before you begin, find out about your bottom's health status as people with HIV, diabetes, and other issues don't heal well. You should be more conscious about that subject as you can also cause open flesh wounds, which are susceptible to disease. I use Technicare, an antic-bacterial surgical prep, to help prevent infections and diseases in many types of scenes. It reduces the likelihood of any infection, and stays active for approximately 24 hours. You can find this at almost any Medical supply store and I highly recommend using this prior to the burn as well as after. Apply it and let it dry before burning. Re-apply after cleaning the burn. I really like Technicare for any type of wound and apply it after a whipping and cigar play as well to help reduce the burning sensation and aid in healing.

A true cigar burn involves the destruction of skin cells, and sometimes the underlying structures of muscle, fascia and rarely but possible bone. It occurs when these structures absorb more heat than they can dissipate. What you do for a burn in the first few minutes after it occurs can make a difference in the severity of the injury! First stop the burning process! Remove any jewelry or clothing that may cause additional burning or access to the burn after the fact. Also think about swelling after the burn, i.e. a Prince Albert or body jewelry may become an issue in treatment. Pour cool water, not cold, over the burned area for at least three to five minutes. Do not put ice on the burn after the burn as it can make the burn worse. Do not put any type of cream, ointment or salve to the open wound. By all means do not put butter on a burn. All of these things can cause infection due to the oil base and can cause the burn to become worse. Upon thorough cleaning of the wound by the cool water, cover the wound with a sterile bandage or clean, non shredding cloth, or bandage. Check for shock and be sure your bottom is warm and not cold. If needed, cover your victim to keep them warm. If you need to seek medical attention do so now. If the injury is minor, let the pain stop and then spray with an antiseptic spray. Recover the wound and allow the wound to crust over, which is known as scabbing. I recommend once the wound has scabbed over and started to shrink, you can apply a product called Scar Gel and there are many types of these products on the market which will help repair the skin and help avoid scarring. If wounds are not healing and appear to be weeping, or smell bad, seek medical attention as these are signs of developing infection.

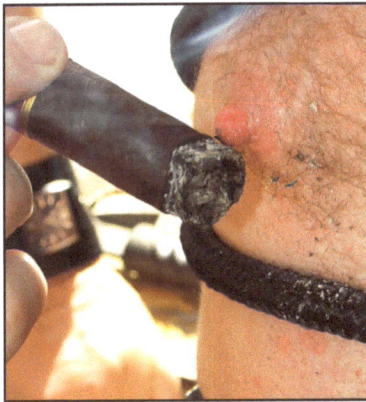

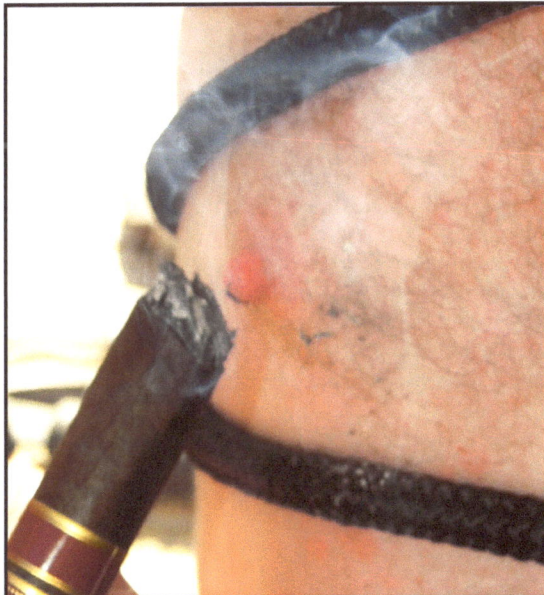

DUMPING ASH IN THE MOUTH/EATING ASH/ ASH IN HAND

Using your bottom's mouth as an ashtray, can also be fun but, has to be done with great precision. The glowing foot of the cigar is where the ash builds. The chunk of glowing ash at the foot of the cigar is known to many as the cherry. The cherry portion or hot ash part can damage the tongue if not removed from the cigar properly. Wait until the ash is about one to one and one half inches long. Have the bottom salivate his tongue heavily. Place the tip edge of the ash on to the bottom's tongue. The spit from your bottom's mouth will grip the ash. You should turn the cigar slowly clockwise; the ash will break off and drop onto the bottom's tongue and mouth. The bottom by heavily salivating will cause less chance of actual burning as the saliva should extinguish the remaining glowing ash. The bottom can then close his mouth and swallow the ash. Please have a bottle of water or some liquid readily available for the bottom should it begin to burn. This will extinguish what remaining glowing ash, which may not be seen, out. This will take some practice and also some patience by both parties.

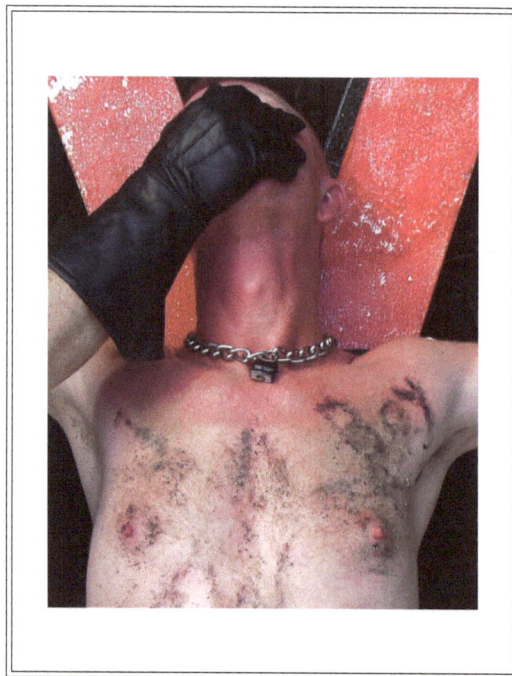

Another practice of a Top and his bottom is to have his bottom hold the hot ashes in his hand. Here the same practice should be done by the Top allowing about one and one half inches of burning ash build before dumping the ash. The top can tap the cigar at this point allowing the ash to fall into the palm of the hand. The palm of the hand can usually handle a little more intense heat should some glowing cherry cigar embers still exist as the palms of the hands have tougher skin. Again, have water available for the bottoms hand should a burning sensation occur. Once this has cleared check for any burns. If some occur treat accordingly.

CLEANING THE LEATHERS AFTER PLAY OR A GOOD SMOKE

Now a day, the smell of smoke is not pleasant to everyone. The dominant, if he so wishes, can have his leather cleaned easily at home after being in a smoke-filled room. This can be done by anyone from the leatherboy, if the Dominant is lucky enough to have one, or he can have someone clean it or he can do the cleaning and maintenance himself. This should be the Dominant's decision, of course, but many people do not enjoy the smell of cigar or cigarette smoke.

To remove the smell of smoke from leather, use undiluted vinegar. Take a spray bottle and put in a teaspoon of baking soda, two tablespoons undiluted vinegar and two cups of water. This creates foam. Once the foaming has stopped, replace the bottle lid and spray the solution as required completely over all the exposed leather. You should spray it directly onto the leather and wipe it off immediately. Then let it air dry COMPLETELY. If the smell still hasn't been removed, put your leather in a plastic bag and put a bowl of vinegar in the bag and seal the bag for about one week. Your leather will come out smelling like vinegar instead of smoke so let it dry completely airing out the vinegar. You will need to condition the leather either way after it has air dried. Condition the leather with Huberd's shoe grease or Dubbin's (a leather conditioner) immediately following the drying process of the leather. Do not do this with suede leather.

IN CONCLUSION

As you see, there are many ways to play with cigars and cigar smoke and ash. It has great panic and scare tactics, it has burning power and it literally is a dangerous play, but safe with an educated top. This type of play should not be done by a beginning top. Should you want to learn, take lessons from a very experienced Master or top. The same goes for bottoms; you need to trust and truly trust your educated top and know that danger is there, which is why many of us like this type of play. I know I can handle my bottom and can revive my bottoms should an emergency occur. I always have a cell phone in my dungeon space for any type of emergency. Luckily in my entire career I have never had to call 911, but I had one time where I thought I was going to have to, as a bottom lied about his condition and didn't tell me what was wrong until after a mishap occurred. Thank God for my skills, as it saved both him and I, his life and my reputation and life.

Don't do any of this type of play until you are proficient and have the skills to handle any situation mentioned. If you're the top you are responsible for safe play and being educated in the play. If you are the bottom, you are responsible for being truthful, honest, and surrendering your trust factor to that top. Question the top to his education, the amount of times he has played. Do you know of this top or Master or is he just someone who is out there proclaiming he knows. Trust me when I say you will find out very quickly once you submit but then it may be too late. Do yourself a favor and negotiate everything up front and see if the top can be trusted to uphold his end of the play. There is nothing wrong with meeting and discussing any type of play. Then plan the play for a later date, so you as the bottom have time to find out if the Master, Sir, or Top has a good reputation. Be sure the top is educated and knows what he is doing. The Top can also find out if you are a good bottom. Both parties play important roles in the negotiating. So please do your part and force the partner to do his part.

Remember, two of the Leathermen's creeds: **RACK**-risk aware consensual kink, and **Safe, Sane and Consensual.** These both apply to Cigar Play! Enjoy and have fun.

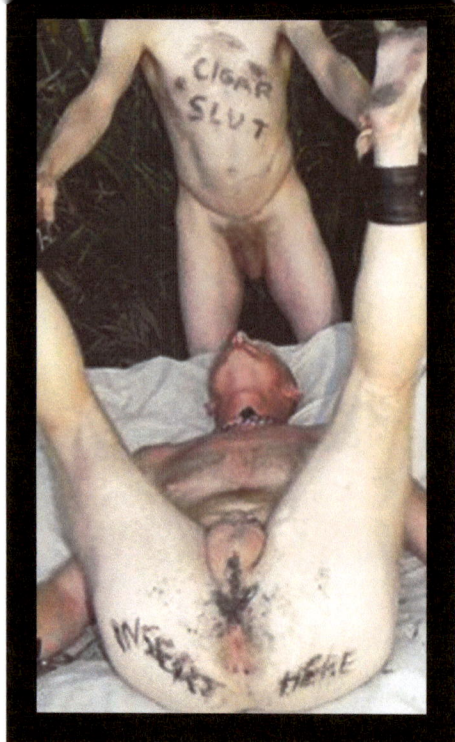

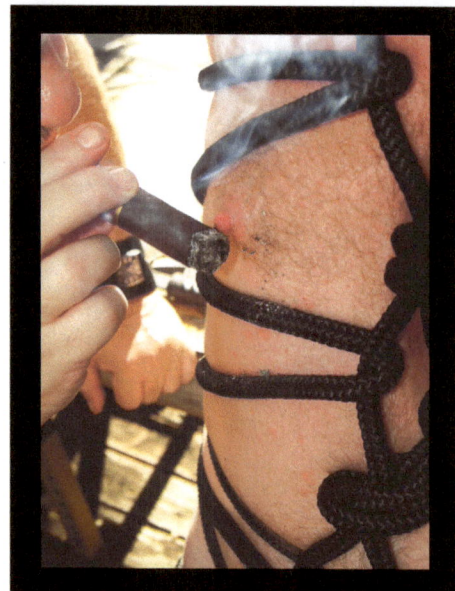

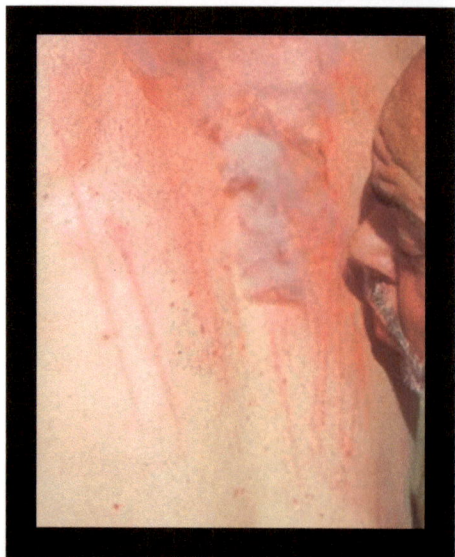

ABOUT THE AUTHOR

Master John has been actively involved with the BDSM Lifestyle for over 35 years. He started serving Master Richard in true Old Guard ways in the Castro in the early 70's.

Master John was Honorary Member of the Year 2008 by the SFLBC. He was a principle in Daddy's Closet located in Wilton Manors, Fl. and hosts the website: www.DungeonAcademy.com. He has worked closely with BDSM educators including Jack Mc George, Robert Dante, Catherine Gross, Sebastian, and Pandora to name a few. He was the publisher and editor of the Leather Link, a publication of leather events, stories etc., in the Ft Lauderdale area. He is a past member of the Florida NLA, Ft Lauderdale, Fl. He is a past sponsor of the ILSb (International Leather Sir boy) contest regionally and in 2009 was a feature theatre presenter and educator for ILSb in San Francisco.

Actively involved in educating our community through Mr. S, LeatherWerks, SFLBC, Leather University, Mens Academy, ILSb, Beyond Leather and S.P.I.C.E., his educational focus embraces both gay and pansexual lifestyles. He is a voting member and contributor to the Leather Archives. He is an experienced judge at many contests globally. Master John co-chairs Leather Title Holders Sunday at the M.C.C. Sunshine Cathedral annually in Ft. Lauderdale. He is involved in charity work at Tuesday's Angels, Center One, The Leather Masked Ball, Lighthouse for the Blind, Care Resources and The Children with Aids Network. In 2009 he became Master Boo to Jeffrey Payne, IML 2009 and Lamalani, IMSL 2009. Over the past five years, he has attended and presented at more than 200 events.

He currently is working on his next book, Old Guard Traditions and Values, featuring many of the past events and traditions which have almost disappeared and not been written about to date.

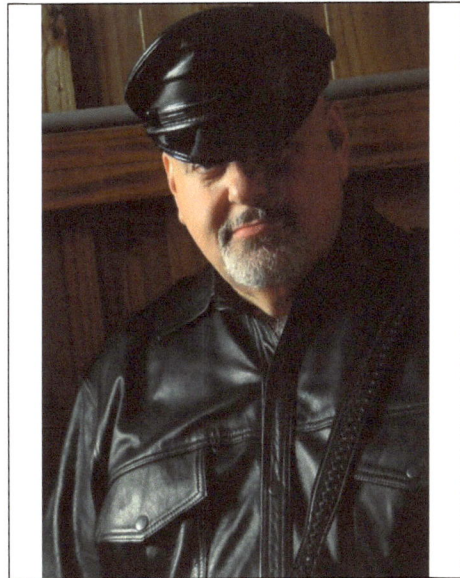

www.ingramcontent.com/pod-product-compliance
Lightning Source LLC
Chambersburg PA
CBHW060818270326
41930CB00002B/81